FUN WITH MATH

[A book for the students preparing for any type of competition]

Lalit Kaushik
B.Tech.(Mechanical); MBA (Consultancy); C.E.(INDIA)

*To My Wife
Priyanka Kaushik*

Preface The First Edition

I am pleased to present the first edition of basic mathematics concept based on Vedic Math. It envelops enough of solved problems for better understanding.

The book comprises 9 Chapters. All chapters are saturated with much needed text supported by simple and self-explanatory figures/symbols.

The entire book has been thoroughly checked and put exercise and test paper for better practicing. Beside this some old questions are also mentioned in last enriched with additional questions drawn from the latest competitive examinations (UPSC, C-Sat etc.) to make it a still more useful.

This book will prove to be boon to the students preparing for UPSC, State services, CTET, UPTET, SSC, DSSB and other competitive examinations.

Although every case has been taken to make the book free from errors both in text as well as in solved examples, yet the author shall feel obliged if errors present are bought to his notice. Any suggestions for the improvement of this book will be thankfully acknowledged and incorporated in the next edition.

AUTHOR

Vedic Math

The term 'Vedic' came from a Sanskrit word 'Veda', that means 'Knowledge'. And, Vedic Math is a super collection of sutras to solve math problems in a faster & easy way.

You can solve many difficult/ time-consuming ICSE/ CBSE /UPSC/SSC/MBA/BBA/etc. Math immediately using Vedic Math Tricks. Moreover, just by using Vedic Math you can solve a problem mentally and that's the beauty of Vedic Maths. While you encounter big digit calculation in competition knowledge of Vedic Math will lend a helping hand to beat the difficulty level of those sums.

We suggest to start initially chapter 9 for better grasping. Chapter 9 has all tools which will be beneficial to be mastery in that.

This is a basic math which will enhance your potential of calculation in simple to complex calculations either in day-to-day activities or in competition to do better than others. After mastery you will relies after mastery that you have some confident in other areas too.

What is Test?

- Test Doesn't mean the scores either low or high.
- Test is the feedback of your true input.
- Enjoy your tests to know about yourself.

Chapter -1

10% Rule

10% rule of any number is the magic key to found out % value of any number. Let us begin with some examples.

Rule - If we want to know 10% of any given number is just divide by 10 to that number.

For example

$$10\% \text{ of } 100 = \frac{100}{10} = 10$$

$$10\% \text{ of } 60 = \frac{60}{10} = 6$$

$$10\% \text{ of } 2,00,000 = \frac{200000}{10} = 20000$$

$$10\% \text{ of } 324 = \frac{324}{10} = 32.4$$

And so, on we can easily find out 10% of any value. Before move further you must do it mastery without paper pencil.

The magic use of this rule is inside their applications.
If we want to know the percentage of any number, that moment we can take advantage of 10% Rule.

For example

(i) 11% of 100

Solve

10% of 100=100/10=10

10% of 10= 10/10 =1

As we know that 10% of 10 is 1% of 100.

Therefore 11% = 10% + 1%

On putting their Values   10 + 1 =11

[11 is the answer]

(ii) 11% of 324

Solve
10% of 324=324/10=32.4

10% of 32.4=32.4/10=3.24

As we know that 10% of 32.4 is 1% of 324.

Therefore 11% = (10% + 1%) of 324

On putting their values 32.4 + 3.24=35.64

[35.64 is the answer]

(iii) 24% of 100

Solve

First, we need to segregate 24%
$$= 2 \times 10\% + 4 \times 1\% \quad \text{--- (1)}$$

10% of 100 = 100/10 = 10

(1% of 100) 10% of 10 = 10/10 = 1

Putting these value in eq^n (1)

$$= 2 \times 10 + 4 \times 1 = 24$$

[24 is the answer]

(iv) 59% of 100

Solve

First, we need to segregate 59%
$$= 6 \times 10\% - 1\% \quad \text{--- (1)}$$

10% of 100 = 100/10 = 10

(1% of 100) 10% of 10 = 10/10 = 1

Putting these values in eq^n (1)

$$= 6 \times 10 - 1 = 59$$

[59 is the answer]

(v) 79 % of 759

Solve
First, we need to segregate 79%
$$= 8 \times 10\% - 1\% \quad \text{--- (1)}$$

10% of 759 = 759/10 = 75.9

(1% of 100) 10% of 75.9 = 75.9/10 = 7.59

Putting these value in eq^n (1)
$$= 8 \times 75.9 - 7.59$$
$$= 607.2 - 7.59$$
$$= 599.61$$

[599.61 is the answer]

(vi) 163 % of 759

Solve
First, we need to segregate 163%
$$= 100\% + 6 \times 10\% + 3 \times 1\% \quad \text{--- (1)}$$

10% of 759 = 759/10 = 75.9

(1% of 100) 10% of 75.9 = 75.9/10 = 7.59

Putting these value in eq^n (1)
$$= 759 + 6 \times 75.9 + 3 \times 7.59$$

$$=759+455.4+22.77$$

$$=1237.17$$

[1237.17 is the answer]

(vii) 9% of 965454

Solve
First, we need to segregate 9%
$$= 10\% - 1\% \text{ --- (1)}$$

10% of 965454 = 965454/10 = 96545.4

(1% of 100) 10% of 96545.4 = 96545.4/10
$$= 9654.54$$

Putting these values in eq^n (1)
$$=96545.4-9654.54$$

$$=86890.86$$

[86890.86 is the answer]

Exercise 1.1

Before start doing exercise, recap and think twice. Have you grasped concept of 10%, if no than move back and learnt it by heart?

1. Break the following on the basis of 10% rule.

1	32%
2	62%
3	69%
4	42%
5	33%
6	72%
7	90%
8	5%
9	9%
10	29%
11	45%
12	49%
13	78%
14	84%
15	86%
16	21%

2. To find the value of following

   a) 10%   Value OF   60
   b) 10%   Value OF   30
   c) 21%   Value OF   100
   d) 22%   Value OF   112
   e) 29%   Value OF   200
   f) 14%   Value OF   32,456
   g) 7%   Value OF   69

Test yourself 1.1　　　　　　　**Time: - 15-20 Min.**

Fill in the blanks.
- a) ___% OF 100 = 10
- b) 1% OF 10 = ____
- c) 3% OF 100 = ____
- d) 7% OF 100 = ____
- e) 14% OF 200 = ____
- f) 14% OF 1000 = ____
- g) 14% OF 700 = ____
- h) 29% OF 300 = ____
- i) 9% OF 724 = ____
- j) 76% OF 396 = ____

Mark yourself.

Chapter – 2

Multiplication through base method (Base 100)

Example 1. 103×104

Step 1: Break into two in ref. of base 100

$$\begin{array}{cc} \text{Left Part} & \text{Right Part} \\ 103 \; + & 3 \\ 104 \; + & 4 \end{array}$$

Right Part: Put number in respect of 100 base, sign always come in respect of that. Here

→In 103, + 3 is more in base value & in 104, + 4 is more in base value.

Left Part: Put value as it is

Step 2: Multiply right-side values.

$$3 \times 4 = 12$$

Step 3: Add cross value in either way.

$$103 + 4 = 107$$
$$Or$$
$$104 + 3 = 107$$

Step 4: Then put right and left side value adjacent to each other than outcome is the answer.

        Left Part   Right Part
          107 : 12

= **10,712 Answer**

Important consideration if right side value come **-ve** or more than two digit you have to accommodate with left outcome. We will explain with example:

Example 2: 96 × 103

Step 1: Break into two in ref. of base **100**

       Left Part   Right Part
         96   −   4
        103   +   3

Right Part: Put number in respect of 100 base, sign always come in respect of that. Here
→In 96, **- 4** is less in base value & in 103, **+ 3** is more in base v*alue*.

Left Part: Put value as it is

Step 2: Multiply right-side values

$$3 \times {}^-4 = {}^-12$$
$$\text{Or}$$
$$-4 \times 3 = {}^-12$$

Step 3: Add cross value in either way

$$96 + 3 = 99$$
$$Or$$
$$103 - 4 = 99$$

Step 4: Then put right and left side value adjacent to each other

Left Part	Right Part
99 :	- 12

Step 5: We have to convert - ve value into + ve value thus borrow from left part

Left Part	+ 1 Right Part
99	− 12
− 1	+100
98	88

Step 6: Then put right and left side value adjacent to each other than outcome is the answer.

Left Part	Right Part
98 :	88

= 9,888 Answer

Example 3 : Where both right side value will come − ve.

97 ×94

Step 1: Break into two in ref. of base **100**

$$
\begin{array}{ccc}
\text{Left Part} & & \text{Right Part} \\
97 & - & 3 \\
\underline{94} & - & \underline{6}
\end{array}
$$

Right Part: Put number in respect of 100 base, sign always come in respect of that. Here

→In 97, **- 3** is less in base value & in 94, **- 6** is less in base value·

Left Part: Put value as it is

Step 2: Multiply right-side values

$$-3 \times -6 = +18$$
$$Or$$
$$-6 \times -3 = +18$$

Step 3 : Add cross value in either way

$$97 - 6 = 91$$
$$Or$$
$$94 - 3 = 91$$

Step 4: Then put right and left side value adjacent to each other than outcome is the answer.

Left Part		Right Part
91	:	18

= 9,118 Answer

Example 4 : Where right side outcome is more than two digit.

$$112 \times 113$$

Step 1: Break into two in ref. of base **100**

Left Part		Right Part
112	+	12
113	+	13

Right Part: Put number in respect of 100 base, sign always come in respect of that. Here

→In 112, **+ 12** is more in base value & in 113, **+ 13** is more in base value.

Left Part: Put value as it is

Step 2 : Multiply right side values

$$12 \times 13 = 156$$
$$Or$$
$$13 \times 12 = 156$$

Step 3 : Add cross value in either way

$$112 + 13 = 125$$
$$Or$$
$$113 + 12 = 125$$

Step 4: Then put right and left side value adjacent to each other than outcome is the answer.

Left Part	Right Part
125 ⋮	156

Step 5: Carry forward third place digit to left part, as per rules two digit is allowed at right side. Thus, we have to carry forward extra towards left side.

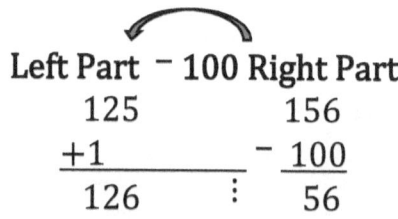

Step 6: Then put left and right-side outcome adjacent to each other than outcome is the answer.

Left Part	Right Part
126 ⋮	56

$$= 12{,}656 \text{ Answer}$$

Example 5: Where right side outcome is more than two digit and right-side value is $-ve$.

$$85 \times 116$$

Step 1: Break into two in ref. of base **100**

	Left Part		Right Part
	85	−	15
	116	+	16

Right Part: Put number in respect of 100 base, sign always come in respect of that. Here

→In 85, **− 15** is less in base value & in 116, **+ 16** is more in base value.

Left Part: Put value as it is

Step 2 : Multiply right side values

$$-15 \times 16 = -240$$
$$Or$$
$$16 \times -15 = -240$$

Step 3: Add cross value in either way

$$85 + 16 = 101$$
$$Or$$
$$116 - 15 = 101$$

Step 4: Then put right and left side value adjacent to each other.

Left Part		Right Part
101	:	− 240

Step 5: We have to convert **- ve** value into **+ ve** value thus borrow from left part

$$
\begin{array}{cc}
\text{Left Part} & \overset{\frown}{+ 3} \text{ Right Part} \\
101 & -\ 240 \\
\underline{-\ 3} & \underline{+\ 300} \\
98 & 60
\end{array}
$$

Step 6: Then put right and left side value adjacent to each other than outcome is the answer.

$$
\begin{array}{ccc}
\text{Left Part} & & \text{Right Part} \\
98 & : & 60
\end{array}
$$

$= 9{,}860$ **Answer**

Exercise 2.1

Solve the following with 100 base methods.

1. 84 X 86
2. 115 X 113
3. 94 X 92
4. 77 X 112
5. 99 X 100
6. 111 X 146
7. 66 X 177
8. 75 X 99
9. 103 X 105
10. 94 X 109
11. 79 X 125
12. 87 X 112

Test Yourself 2.1 **Time : 10 Min.**

1. 81 × 90
2. 99 × 98
3. 125 × 136
4. 188 × 177
5. 111 × 112
6. 88 × 89
7. 76 × 75
8. 87 × 84
9. 112 × 116
10. 125 × 88

Note:
- Try to do all calculation within a time frame.

- Mark Yourself.

- If not scored 100% then revise your basics once again.

Cherish every moment, no one can steal from you, until you give them a chance

Chapter – 3

Multiplication through base method (Base 200) when base is not equal to 100

Base : 200

Example.1 208 × 212

Step 1: Break into two in ref. of base **200**

	Left Part	Right Part
	208 +	8
	212 +	12

Right Part: Put number in respect of 200 base, sign always come in respect of that. Here

→In 208, **+8** is more in base value & in 212, **+12** is more in base value.

Left Part : Put value as it is

Step 2 : Multiply right side values

$$12 \times 8 = 96$$
Or
$$8 \times 12 = 96$$

Step 3 : Add cross value in either way

$$208 + 12 = 220$$
Or
$$212 + 8 = 220$$

Step 4: Then put right and left side value adjacent to each other.

Left Part	Right Part
220	96

Step 5: Multiply 2 in left part outcome and that will be the left part final out come
$$2 \times 220 = 440$$

Step 6: Then put value adjacent to each other than outcome is the answer.

Left Part	Right Part
440	96

$$= 44,096 \text{ Answer}$$

Example.2 When right side value -ve.
$$196 \times 215$$

Step 1: Break into two in ref. of base **200**

Left Part		Right Part
196	−	4
215	+	15

Right Part: Put number in respect of 200 base, sign always come in respect of that. Here

→In 196, **-4** is less in base value & in 215, **+15** is more in base value.

Left Part: Put value as it is

Step 2 : Multiply right side values

$$15 \times -4 = -60$$
$$Or$$
$$-4 \times 15 = -60$$

Step 3 : Add cross value in either way

$$196 + 15 = 211$$
$$Or$$
$$215 - 4 = 211$$

Step 4: Then put right and left side value adjacent to each other.

Left Part	Right Part
211	-60

Step 5: Multiply 2 in left part outcome and that will be the left part final out come

$$2 \times 211 = 422$$

Step 6: Then put value adjacent to each other.

Left Part	Right Part
422	-60

Step 7: We have to convert **-v** value into **+ve** value thus borrow from left part.

$$\text{Left Part} \quad + 1 \quad \text{Right Part}$$
$$\begin{array}{rr} 422 & -\ \ 60 \\ -\ 1 & +100 \\ \hline 421 & 40 \end{array}$$

Step 8: Then put value adjacent to each other than outcome is the answer.

Left Part	Right Part
421	40

= **42,140 Answer**

> **Flyover law:** When we borrow from left part it value is 100 times to right part. Similarly, when left borrow from right it value will be 100th part of left part. Condition applies.

Example.3 When right side multiplication is more than double digit

$$215 \times 214$$

Step 1: Break into two in ref. of base **200**

Left Part	Right Part
215 +	15
214 +	14

Right Part: Put number in respect of 200 base, sign always come in respect of that. Here
→In 215, **+15** is more in base value & in 214, **+14** is more in base value.

Left Part : Put value as it is

Step 2 : Multiply right side values

$$15 \times 14 = 210$$
$$\text{Or}$$
$$14 \times 15 = 210$$

Step 3 : Add cross value in either way

$$215 + 14 = 229$$
$$\text{Or}$$
$$214 + 15 = 229$$

Step 4: Then put right and left side value adjacent to each other.

Left Part	Right Part
229 :	210

Step 5: Multiply 2 in left part outcome and that will be the left part final out come
$$2 \times 229 = 458$$
(Here 2 comes from base value **200**)

Step 6: Then put value adjacent to each other

Left Part	Right Part
458 :	210

Step 7: Right side outcome is 3 digits as per rule two digit is allowed at right side. Thus, we have to carry forward extra towards left side.

$$
\begin{array}{cc}
\text{Left Part} \quad +\ 200\ \text{Right Part} \\
458 \qquad\qquad 210 \\
+\ 2 \qquad\qquad -\ 200 \\
\hline
460 \qquad\qquad 10
\end{array}
$$

Step 8: Then put value adjacent to each other than outcome is the answer.

$$
\begin{array}{cc}
\text{Left Part} & \text{Right Part} \\
460 & 10
\end{array}
$$

$$= 46{,}010 \text{ Answer}$$

When base value 300 to 900

Rule:- Rule will be same as earlier discussion. Only difference is this we multiply right side value with **2** here we multiply **3** when base is **300**, **4** when base is **400**, **5** when base if **500**, **6** when base is **600**, **7** when base is **700**, **8** when base is **800**, **9** when base is **900**.

Example.4 When base value is **300**

$$253 \times 301$$

$$
\begin{array}{cc}
\text{Left Part} & \text{Right Part} \\
253 & -\ 47
\end{array}
$$

$$\frac{301}{254} \qquad \frac{+1}{-47}$$

Step -   3 (254)   − 47
Step -   762   − 47
Step -   − 1   +100
Step -   761   53

=76,153 Answer

Explanation

1. Put right part value as per base value. Here in 253, − 47 is right side value as − 47 less in base value. 301, + 1 is right side value as + 1 is extra in base value.

2. Multiply right-side value
$$(-47) \times (+1) = -47$$

3. Add cross value in either way and put outcome at left part
$$253 + 1 = 254$$
or
$$301 - 47 = 254$$

4. Multiply left side value with 3
$$3 \times 254 = 762$$

(Here 3 comes from base value)

5. We have to convert **-ve** value into **+ve** value thus borrow from left side outcome.

$$\begin{array}{cc} \text{Left Part} \;_{+\,1}\; \text{Right Part} \\ 762 & -47 \\ -1 & +100 \\ \hline 761 & 53 \end{array}$$

6. Therefore, then put left and right-side outcome adjacent to each other than outcome is the answer.

$$\begin{array}{cc} \text{Left Part} & \text{Right Part} \\ 761 & \vdots \quad 53 \end{array}$$

$$= 76{,}153 \text{ Answer}$$

Example.5 When base value is **900**

$$876 \times 911$$

$$\begin{array}{cc} \text{Left Part} & \text{Right Part} \\ 876 & -24 \\ 911 & +11 \\ \hline 887 & -264 \end{array}$$

Step -	9 (887)	−264
Step -	7983	−264
Step -	-3	+300
Step -	7980	36

$$= 7{,}98{,}036 \text{ Answer}$$

Explanation

1. Put all values according to rule of base at right side. Here in $876, -24$ is right side value as -24 less in base value. $911, +11$ is right side value as $+11$ is extra in base value.

2. Multiply right-side value

$$(-24) \times (+11) = -264$$

3. Add cross value in either way and put outcome at left part

$$876+11=887$$
$$Or$$
$$911-24=887$$

4. Multiply left side value with 9

$$9 \times 887 = 7,983$$

(Here 9 comes from base value 900)

5. We have to convert **-ve** value into **+ve** value thus borrow from left side outcome.

```
Left Part + 3Right Part
7983          300
 - 3         - 264
─────        ─────
7980           36
```

6. Therefore, then put left and right-side outcome adjacent to each other than outcome is the answer.

Left Part   Right Part
7980   ׃   36

= 7,98,036 Answer

Exercise 3.1

Predict nearest base value and multiply with the help of base value.

1. 101 X 102
2. 98 X 97
3. 103 X 112
4. 210 X 195
5. 195 X 196
6. 115 X 109
7. 703 X 650
8. 525 X 495
9. 900 X 825
10. 201 X 98
11. 233 X 201
12. 108 X 97
13. 160 X 210
14. 333 X 217
15. 536 X 595
16. 444 X 425
17. 636 X 701
18. 204 X 203
19. 601 X 646
20. 909 X 876

Test Yourself 3.1 Time: 20-30 Min

Predict nearest base value and multiply with the help of base value

1	97	X	94
2	87	X	94
3	63	X	94
4	106	X	105
5	111	X	96
6	201	X	190
7	187	X	211
8	237	X	301
9	266	X	276
10	330	X	340
11	11.404	X	480
12	12.467	X	503
13	13.607	X	675
14	14.701	X	776
15	15.787	X	804
16	16.603	X	690
17	17.801	X	811
18	18.901	X	906
19	19.333	X	296
20	20.777	X	801

Note: -

1. Each question is for 1 mark and 1 minutes in maximum time. If you complete within 20 minutes then it is ok and if you finished beyond that within 30 minutes then you have to brush up earlier concept chapter once again.

2. If you complete this task within 10 minutes, you are good in practice and if you finished without error within five minutes then you are genius.

3. Number score
        20/20-Wow
        (15-19)/20-Good
        (10-14)/20-Poor
        (1-9)/20-Very Poor

4. Practice with accuracy is the key of wow marks.

Chapter – 4

Condition: - Multiplication when sum of unit digit is 10 and rest of digit are same or similar.

Example 1: - 7 2 × 7 8

Step a: In this first check condition sum of unit digit is 10 and rest of digit are same or similar.

$$72 \times 78$$
$$2 + 8 = 10$$

[rest of digit are same or similar]

Step b: This condition satisfies. Now move on towards calculation

Left Part : Right Part
72   :   + 2
78   :   + 8
7(7 + 1):8 × 2

56:16

Step c: Therefore, then put left and right-side outcome adjacent to each other than outcome is the answer.

                    Left Part   Right Part
                       56   ⋮ 16

                                        = 5,616 Answer

Rule: In the multiplication when unit place digits sum is 10 and rest of the digit are same or similar. Then, this rule is applicable. In this put unit place digit at right part and normal input number at left part. Multiply unit place digit and multiply next place digit with next consecutive number. Then put both outcome adjacent to each other, then outcome is the answer.

Symbolic Formula: - If given digits are
$$AB \times AC \quad \text{Where } B + C = 10$$

Then $[A(A+1) \vdots B \times C]$
$$72 \times 78, [Whereas\ A = 7, B = 2, C = 8]$$
$$= 5,616\ \text{Answer}$$

Explanation

1. Put right side unit place digit

2. Multiplying right side values

   $2 \times 8 = 16$

3. Multiply next to unit place digits or value with next consecutive number.

   $7 \times (7+1) = 7 \times 8 = 56$

4. Then put left and right outcome adjacent to each other than outcome is the answer

$$\textbf{Left outcome} : \textbf{Right outcome}$$
$$56 \qquad : \qquad 16$$

$$= 5{,}616 \text{ Answer}$$

Example 2: 117 × 113

Step a: Check Condition
- Unit place number 7 & 3

   Both sum 7 + 3 = 10 →Satisfy

- Next digit number 11 & 11

   Both are same →Satisfy

Thus, rule condition satisfies

Step b: Put value left and right part

$$\begin{array}{r|l}
\textbf{Left Part} & \textbf{Right Part} \\
117 & 7 \\
113 & 3 \\ \hline
11 \times 12 & 7 \times 3
\end{array}$$

$$132{:}21$$

Step c: Therefore, then put left and right-side outcome adjacent to each other than outcome is the answer.

Left Part		Right Part
132	:	21

= 13,221 Answer

Explanation

1. Put right side unit place digit. Hence digits are 7 & 3

1. Put left side value as it is. Here values are 117 & 113

2. Multiplying right side values

   $7 \times 3 = 21$

3. Multiply next to unit place digit to next consecutive number

   $11 \times 12 = 132$

4. Then put left and right outcome adjacent to each other than outcome is the answer

Left outcome		Right outcome
132	:	21

= 13,221 Answer

Example 3:   506 × 504

Step a: Check Condition

- Unit place number 6 & 4
  Both sum $6 + 4 = 10 \rightarrow$ Satisfy

- Next digit number 50 & 50
  Both are same $\rightarrow$ Satisfy

Thus, rule condition satisfies

$$
\begin{array}{c|c}
\text{Left Part} & \text{Right Part} \\
506 & 6 \\
\underline{504} & \underline{4} \\
50 \times 51 & 6 \times 4 \\
\end{array}
$$

$$2550 : 24$$

$$= 2,55,024 \text{ Answer}$$

Explanation

1. Put right side unit place digit. Here 6 & 4 one-unit place digit.

2. Put left side value as it is. Here values are 506 and 504.

3. Multiplying right side value

$$6 \times 4 = 24$$

4. Multiply next to unit place digit to next to consecutive number

$$50 \times 51 = 2550$$

5. Then put left and right outcome adjacent to each other, then outcome is the answer.

Left outcome	Right outcome
2550	24

= 2,55,024 Answer

Example 4: 1996 × 1994

Step a: Check Condition

- Unit place number 6 & 4
    Both sum 6 + 4 = 10 →Satisfy

- Next digit number 199 & 199
    Both are same →Satisfy

Thus, rule condition satisfies

Left Part	Right Part
1996	6
1994	4
199 × 200	6 × 4
39800	24

= 39,80,024 Answer

Explanation

1. Put right side unit place digit. Here 6 & 4 one-unit place digit.

2. Put left side value as it is. Here values are 1996 and 1994.

3. Multiplying right side value

    6 × 4 = 24

4. Multiply next to unit place digit to next to consecutive number

    199 × 200 = 39800

5. Then put left and right outcome adjacent to each other, then outcome is the answer.

    Left outcome   **Right outcome**
    39800   ⋮   24

    = 39,80,024 Answer

Exercise 4.1

Kindly check first rule condition and solve the multiplication.

1. 306 X 304
2. 201 X 209
3. 334 X 336
4. 226 X 224
5. 111 X 119
6. 24 X 26
7. 32 X 38
8. 43 X 47
9. 54 X 56
10. 77 X 73
11. 93 X 98
12. 78 X 72
13. 102 X 108
14. 36 X 34
15. 1112 X 1118
16. 83 X 87
17. 99 X 91
18. 62 X 68
19. 509 X 501
20. 251 X 259

Test Yourself 4.1   Time: 10-15 Min.

Kindly satisfy first rule conditions and solve the multiplications.

1. 32 X 38
2. 101 X 109
3. 709 X 701
4. 22 X 28
5. 603 X 607
6. 604 X 607
7. 303 X 307
8. 94 X 98
9. 401 X 409
10. 606 X 604

Note:

1. Each question is for 1 mark and 1 minute is the maximum time. If you complete this test task within 10 minutes then it is ok and if you finished beyond that within 15 minutes. Then you have to brush up earlier chapter once again.

2. If you complete this task within 5 minutes, you are good in practice and if you finish this test task without error within 2.5 minutes then you are genius.

3. Number score
   10/10-Wow

(7-9)/10-Good

(5-6)/10-Poor

(1-4)/10-Very Poor

4.   Lower mark does not mean; you are weak in math. It's meant your focused is disturbing. Work on it.

Chapter – 5

Condition: - Application for multiplication, when unit place digit is **5** and rest of the place digits are consecutive number.

MEMORABLE FORMULA FOR THESE CONDITIONS.

                    Left Part : Right Part
                          SOBN -1 : 75

SOBN: Square of bigger number

Example 1: 55 × 65

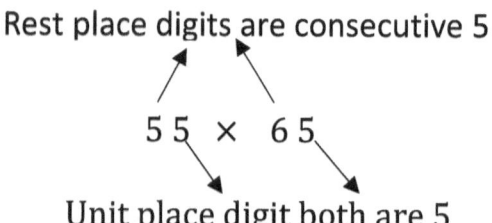

Thus, condition satisfies.

Rule:

1. First Satisfy condition, then move on.

2. In this there are two outcomes, left outcome & right outcome.

3. Right outcome is fixed in this rule, which is 75.

4. For left outcome choose big one consecutive number. Here is 6 is bigger than 5. Thus 6 is selected for further calculation.

5. Square of this selected number and subtract one from it.

$$6^2 - 1 = 36 - 1 = 35$$

6. Then put left and right outcome adjacent to each other. Then outcome is the answer.

Left Part Outcome	:	Right Part Outcome
35	:	75

= 3575 Answer

Example 2: 115×125

- Here unit place digits are same and both are 5, Thus satisfied the condition.

- Rest of the place digit are 11 & 12 and both are consecutive number, thus satisfied the condition.

Step a: After satisfaction of condition, put bigger no after unit place digit 12 in formula and solve step by step.

$$115 \times 125$$
$$12^2 - 1 : 75$$

$$144 - 1 : 75$$

$$143 : 75$$

Step b: Then put left and right outcome adjacent to each other. Then outcome is the answer.

Left Part : Right Part
Outcome : Outcome
143 :   75

= 14,375 Answer

Explanation

1. First satisfied the conditions.

2. Right side outcome is fixed, which is 75 in this rule.

3. For left outcome, choose big number and square of it then subtract 1 from that outcome
$$12^2 - 1 = 144 - 1 = 143$$

4. Then put left and right outcome adjacent to each other then outcome is the answer.

                    Left outcome : Right outcome

                         143   :   75

                                     = 14,375 Answer

Example 3:   3115 × 3125

⊥ Here unit place digits are same and both are 5, Thus satisfied the condition.

⊥ Rest of the place digit are 311 & 312 and both are consecutive number, thus satisfied the condition.

Step a: After satisfaction of condition, put bigger no after unit place digit 12 in formula and solve step by step.

$$3115$$
$$\times 3125$$
$$312^2 - 1 : 75$$

$$97344 - 1 : 75$$

$$97343 : 75$$

Step b: Then put left and right outcome adjacent to each other. Then outcome is the answer.

Left Part : Right Part
Outcome : Outcome
97343 : 75

= 97,34,375 Answer

Explanation

1. First satisfied the conditions.

2. Right side outcome is fixed, which is 75 in this rule.

3. For left outcome, choose big number and square of it then subtract 1 from that outcome

$$312^2 - 1 = 97344 - 1 = 97343$$

4. Then put left and right outcome adjacent to each other than outcome is the answer.

Left outcome	:	Right outcome
97343	:	75

= 9734,375 Answer

Exercise 5.1

Kindly check first rule condition and solve the multiplication.

1. 65 X 75
2. 85 X 95
3. 35 X 45
4. 25 X 35
5. 105 X 115
6. 125 X 135
7. 135 X 145
8. 225 X 235
9. 325 X 335
10. 445 X 455

First, they ignore you, then they laugh at you, then they fight you, then you win.

Test Yourself 5.1   Time: 5-10 Min.

Kindly satisfy first this chapter conditions and solve the multiplications. If not satisfy mark not satisfy (N/S)

1.   53   X   55
2.   45   X   55
3.   325   X   315
4.   615   X   625
5.   205   X   215
6.   605   X   615
7.   95   X   105
8.   101   X   105
9.   105   X   115
10   65   X   55

Note:

1. Each question is for 1 mark and 1 minute is the maximum time. If you complete this test task within 10 minutes then it is ok and if you finished beyond that within 15 minutes. Then you have to brush up earlier chapters once again.

2. If you complete this task within 5 minutes, you are good in practice and if you finish this test task without error within 2.5 minutes then you are genius.

3. Number score
   a. $\dfrac{10}{10}$ – Wow
   b. $\dfrac{7-9}{10}$ – Good
   c. $\dfrac{5-6}{10}$ – Poor
   d. $\dfrac{1-4}{10}$ – Very Poor

Winning is not a habit, it is a passion. Work on it with honesty.

Chapter – 6

GENERAL MULTIPLICATIONS
(CRISS – CROSS METHOD)

This criss – cross method is applicable for any two numbers to be multiplied and should be used when no one of the earlier chapter's conditions matches. We are saying so, it is little bit time consuming as compare to earlier define methods. However, result always same. No question which method you are using for solving the problem.

Rule Symbol to memorize.

Rule 1 : For multiplication of two digit number

$$ab \times cd$$

Symbol
$$\begin{array}{cc} a & b \\ \times c & d \end{array}$$

Answer

$$a \updownarrow c \qquad \begin{array}{cc} a & b \\ \times & \\ c & d \end{array} \qquad b \updownarrow d$$

Rule 2 : For multiplication of three digit number

$$abc \times def$$

Symbol   a b c
         ×d e f
         Answer

a	a	b	a	b	c	b	c	c
↕	⤫		⤫			⤫		↕
d	d	e	d	e	f	e	f	f

Example 1. Multiplication of two digit
        62 × 23

Step -1: Put Values vertically
        6   2
        2   3

Step - 2:
      2
   × 3
(Multiply unit digits)
      6

Step - 3
6   2
2   3
18 + 4 = 22
(Cross multiply and add outcome)

Step - 4
    6
 × 2
  12
(Multiply tens place digits)

Step - 5

(Solve by carry forwarding extra digit to next outcome for finalization)

$$12 \quad 22 \quad 6$$

$$= 12 \quad 22 \quad 6$$
$$\quad\quad\text{Fixed Fixed}$$

$$= (12+2) \quad 26 \quad \text{(Add carry forward number to existing one)}$$

$$= \textbf{1,426 Answer}$$
(Outcome is the answer)

Example 2.   Multiplication of two digit
$$53 \times 54$$

Step -1   (Put Values vertically)

$$\begin{array}{cc} 5 & 3 \\ 5 & 4 \end{array}$$

Step - 2

$$\begin{array}{r} 3 \\ \times\ 4 \\ \hline 12 \end{array}$$

(Multiply unit digits)

Step - 3
$$\begin{array}{cc} 5 & 3 \\ 5 & 4 \end{array}$$

(Cross multiply and add outcome)
$$20 + 15 = 35$$

Step - 4

(Multiply tens place digits)

Step – 5

           25   35   12
(Put all values of outcome below symbol)

           25   35   12
(Solve by carry forwarding extra digit to next outcome for finalization)

           25   (35+1)   2

           25   36   2

        (25+3)   6   2
(Adding carry forward number)

           28   6   2
                Fixed

(When all values fixed then put adjacent to each other. Then outcome is the answer

                    = 2,862 *Answer*

(Outcome is the answer)

<u>Example 3.</u>   Multiplication of three digit
               246 × 658

Step -1   2 4 6
                6 5 8

Step - 2   (Put Values vertically)
                        6
                      × 8
                      ──
                      48

Step - 3   4   6
                   5   8
             (Cross multiply and add outcome)
                   32 + 30 = 62

Step - 4   2 4 6
                  6 5 8
             16 + 20 + 36 = 72
             (Cross multiply and add outcome)

Step – 5   2   4
                   6   5
             (Cross multiply and add outcome)
                   10 + 24 = 34

Step - 6   2
                  × 6   (Multiply)
                  ──
                  12

Step - 7

a	a	b	a	b	c	b	c	c
d	d	e	d	e	f	e	f	f

(Solve by carry forwarding extra digit to next outcome for finalization)

12 34 72 62 48

12 34 72 (62+4) 8←Fixed

12 34 72 66 8

12 34 (72+6) 68

12 34 78 68

12 (34+7) 868

12 41 868

(12+4) 1868 ←Fixed

(When all values fixed then put adjacent to each other. Then outcome is the answer)

= 1,61,868 Answer

Exercise 6.1

Kindly check first rule condition and solve the multiplication.

1. 34 X 36
2. 69 X 73
3. 296 X 195
4. 798 X 999
5. 300 X 800
6. 145 X 256
7. 49 X 34
8. 294 X 88
9. 325 X 569
10. 425 X 324

Chapter – 7

SQUARING:

Condition no. 1: - When number is close to 10^n or base values 100, or multiple like 200, 300 .... 900.

Condition no. 2: - When number is close to 50.

<u>In Brief</u>   Condition no. 1
FORMULA FOR MEMORISING WHEN BASE VALUE 100

$$\text{Left Part : Right Part}$$

$$B + 2A : A^2 \quad \rightarrow (i)$$

Example: 1 107^2

B = 100 ; A = 7

$107^2 \rightarrow (100 + 2 \times 7) : 7^2$

Step → 114 : 49

= 11,449 Answer

Explanation

Step 1. Split into two parts. Here in this 100 + 7 whereas B is 100 and 7 is A

Step 2. Put B & A values in memorize Formula.

$$\underset{\downarrow}{B} \quad \underset{\downarrow}{A} \quad \underset{\downarrow}{A}$$
$$(100 + 2 \times 7) : 7^2$$

Step 3. Solve right and left part

$$(100 + 2 \times 7) : 7^2$$

$$(100 + 14) : 49$$

$$114 : 49$$

Step 4. Therefore, then put left and right outcome adjacent to each other than outcome is the answer.

Left part : Right Part
114 : 49

= 11,449 Answer

Example 2. 96^2

$$(100 - 4 = 96)$$
$$B = 100$$
$$A = -4$$

Step a) : $B + 2A \mid A^2$

Step b): $[100 + 2 \times (-4)] : (-4)^2$

Step c) : $(100 - 8) \mid 16$

$$92:16$$

$$= 9{,}216 \; Answer$$

Explanation

Step 1. Split into two parts respects to 100. Here in $100 - 4$

Step 2. Put B and A value in rule formula.

$$B + 2A \mid A^2$$

$$\text{Left part} : \text{Right Part}$$
$$[100 + 2 \times (-4)]:(-4)^2$$

Step 3. Solve right and left part

$$[100 + 2 \times (-4)]:(-4)^2$$

$$100 - 8 : 16$$

$$92 : 16$$

Step 4. Therefore, then put left and right outcome adjacent to each other than outcome is the answer

$$92 : 16$$

$$= 9{,}216 \; Answer$$

FORMULA FOR MEMORISING WHEN BASE VALUE 200
Left Part : Right Part

$$2(B + 2A) : A^2 \to (i)$$

2 comes from base value.

Example 3: When base value 200

$$213^2$$

$$\underset{200}{B} + \underset{13}{A}$$

Step a: $2(200 + 2 \times 13) : 13^2$

Step b: $2(200 + 26) : 169$

Step c: $2(226) : 169$

Step d: $452 \mid 169$

Step e: $453 : 69$

$$= 45{,}369 \text{ Answer}$$

Explanation

1. Split into two parts. Here in 200 + 13. Whereas B is 200 and *A* is 13.

2. Put A & B values in formula

     B   A A
     ↓   ↓ ↓
$$(200 + 2 \times 13) : 13^2$$

3. Solve right and left part

$$200 + 26 : 13^2$$

$$226 : 169$$

4. Multiply left part by 2. Here this two come from base.

$$2 \times 226 : 169$$

$$452 : 169$$

5. Carry forward 1 from right part to left part, As right part has two digit limitation

Left part : Right Part

$$452 : 169$$

$$452 + 1 : 69$$

$$453 : 69$$

6. Then put left and right outcome adjacent to each other than outcome is the answer.

$$453 : 69$$

$$= 45{,}369 \text{ Answer}$$

Note: - Similarly we can calculate any square whether 300 base or 900.

Example 4: 926^2

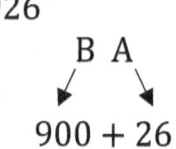

$$900 + 26$$

Left part	Right Part
9 (B + 2 A)	A^2
9 comes from base value	

Step a: $9(900 + 2 \times 26) \mid 26^2$

Step b: $9(900 + 52) \mid 676$

Step c: $9(952) \mid 676$

Step d: $856\overset{\frown}{8} \mid 676$

Step e: $(8568 + 6) \mid 76$

Step f: $8574 \mid 76$

$$= 857476 \text{ Answer}$$

<u>Explanation</u>
1. Split into two parts. Here in 900 + 26. Whereas B is 900 and A is 26.

2. Put *A* & *B* values in formula

$$(900 + 2 \times 26) \mid 26^2$$
(B) (A)

3. Solve right and left part

$$(900 + 2 \times 26) : 26^2$$

$$952 : 26^2$$

4. Multiply left part by 9.
$$9 \times 952 : 676$$

$$8568 : 676$$

5. Carry forward 6 from right part to left part, as right part has constraint of two digit.

Left part : Right Part

$$8568 : 676$$

$$8568 + 6 : 76$$

$$8574 : 76$$

6. Then put left and right outcome adjacent to each other than outcome is the answer.
$$8574 : 76$$

$$= 857476 \text{ Answer}$$

Condition 2 in brief: - When number is close to 50

Rule: -Split into two parts in reference of 50. Put values in formula and get the answer.

[Memorizing formula (25 + A) | A²]

Example 2.1   54²

                50+4 ← (A)

Formula : (25 + A) | A²

Step a)   (25 + 4) | (4)²

Step b)   29 :16

                                    =2916 Answer

Explanation

1. Split into two parts in respect of 50. Here is 50 + 4. Whereas 4 is A.

2. Put A Value in formula.

            Left Part : Right Part
            (25 + A) : A²

3. Solve right and left part
            25 + 4 : 4²

                29 : 16
4. Then put left and right outcome adjacent to each other than outcome is the answer.

$$29 \mid 16$$

$$=2916 \text{ Answer}$$

Example 2.2 32^2

$$50 - \boxed{18} \rightarrow A$$

Here A = - 18

Left part : Right part
Formula : $(25 + A) : A^2$

Step a): $(25 - 18) : (-18)^2$

Step b):   7 : 324

Step c): $7 + 3 : 24$

$$=1024 \text{ Answer}$$

Explanation

1. Split into two parts in respect of 50. Here is $50 - 18$. Whereas A is -18.

2. Put A Value in formula.

Left Part : Right Part
$(25 + A)$: A^2

3. Solve right and left part.
$25 - 18 : (-18)^2$

7:324

4. Carry forward 3 toward left side due to two-digit constraint.

$$\overset{\frown}{7 : 324}$$

$$7 + 3 : 24$$

$$10 : 24$$

5. Then put left and right outcome adjacent to each other than outcome is the answer.

$$10 : 24$$

=1,024 Answer

Exercise 7.1

Kindly check first rule condition and solve the multiplication.

1. 103^2
2. 136^2
3. 211^2
4. 103^2
5. 410^2
6. 625^2
7. 731^2
8. 111^2
9. 98^2
10. 36^2
11. 25^2
12. 77^2
13. 90^2
14. 60^2
15. 176^2
16. 545^2
17. 118^2
18. 671^2
19. 981^2
20. 825^2

Test Yourself 7.1   Time: 20-30 Min

Predict the condition and find out the square.

1. 51^2
2. 63^2
3. 34^2
4. 79^2
5. 116^2
6. 310^2
7. 436^2
8. 749^2
9. 636^2
10. 241^2
11. 119^2
12. 349^2
13. 736^2
14. 824^2
15. 951^2
16. 333^2
17. 42^2
18. 109^2
19. 202^2
20. 88^2

Note:
1. Each question is for 1 mark and 1 minute is the maximum time. If you complete this test task within 20 minutes then it is ok and if you finished beyond that within 30 minutes. Then you have to brush up basics.

2. If you complete this task within 10 minutes, you are good in practice and if you finish this test task without error within 5 minutes then you are genius.

3. Number score

$$\frac{20}{20} - \text{Wow}$$

$$\frac{14\text{-}19}{20} - \text{Good}$$

$$\frac{10\text{-}13}{20} - \text{Poor}$$

$$\frac{1-9}{20} - \text{Very Poor}$$

4. Improve your focus if you are getting less then wow marks.

Chapter – 8

CUBING: We can find the cube of any number close to a power of 10 say 10^n

Formula for memorizing when base value 10 or 10^n

"Left Part : Right Part 2 : Right Part 1"

$$B + 3A : 3 \times A^2 : A^3 \rightarrow (i)$$

Example 1: 109^3

$$\underset{\downarrow}{B} \quad \underset{\downarrow}{A} \quad [B \rightarrow 100, A \rightarrow 9]$$

$$100 + 9$$

Step a: Putting B and A value in formula (i)

$$B + 3A : 3 \times A^2 : A^3 \rightarrow (i)$$

$$100 + 3 \times 9 : 3 \times 9^2 : 9^3$$

Step b: Solve it

$$100+27 : 3 \times 81 : 729$$

$$127 : 243 : 729$$

Step c: Carry forward extra digit from right part to left part as limitation of two digit

$$127:243:729$$

$$127:243+7:29$$

$$127:250:29$$

$$127+2:50:29$$

$$129:50:29$$

Step d: Therefore, then put left and right outcome adjacent to each other than outcome is the answer

$$129:50:29$$

$$=12{,}95{,}029 \text{ Answer}$$

Formula for memorizing when base value 200

"Left Part : Right Part 2 : Right Part 1"

$$2^2(B+3A) : 2 \times 3A^2 : A^3 \rightarrow (ii)$$

Example 2: 196^3

$$\underset{\downarrow}{B} \quad \underset{\downarrow}{A} \quad [B = 200, A = -4]$$

$$200 - \quad 4$$

Step a: Putting B and A value in formula (ii)

$$2^2(B+3A) : 2 \times 3A^2 : A^3 \rightarrow (ii)$$

$$2^2(200 + 3 \times -4) : 2 \times 3(-4)^2 : (-4)^3$$

Step b: Solve it

$$2^2(200 + 3 \times -4) : 2 \times 3(-4)^2 : (-4)^3$$

$$4(200-12) : 6 \times 16 : -64$$

$$4 \times 188 : 96 : -64$$

$$752 : 96 : -64$$

Step c: Remove -ve sign from right part 1 by borrow from right part 2

$$4 \times 188 : 96 : \overset{\frown}{-}64$$

$$4 \times 188 : 95 : 100-64$$

$$4 \times 188 : 95 : 36$$

$$752 : 95 : 36$$

Step d: Therefore, then put left and right outcome adjacent to each other than outcome is the answer

$$752 : 95 : 36$$

$$= 75,29,536 \text{ Answer}$$

Note: - We can do cubing of any number, but from competition point of view up to given level is sufficient

Exercise 8.1

Kindly check first rule condition and solve the multiplication.

1. 96^3
2. 103^3
3. 111^3
4. 201^3
5. 166^3

6. 212^3
7. 233^3
8. 92^3
9. 88^3
10. 142^3

Test Yourself 8.1   Time : 20 Min

1. Cube of 86^3

   a) 636156   b) 636056   c) 736056"

2. Cube of 176^3

   a) 5451778   b) 5451779   c) 5451776

3. Cube of 212^3

   a) 9528328   b) 9528228   c) 9528128

4. Cube 56^3

   a) 175616   b) 175617   c) 175618

5. Cube of 211^3

   a) 9393931   b) 393931   c) 392932

Note: -
1. Before doing this test, kindly revise all earlier chapter once again especially chapter 9

2. Make rating to yourself and improve if necessary

Chapter – 9

MATH WARRIOR TOOLS

Before start one would have to learn the following basic tools of mathematics by heart.

Tables   – **Up to 30th**

Squares   – **Up to 40th**

Cubes   **-Up to 25th**

Square roots   – **Up to 20th**

Cube roots   – **Up to 20th**

Reciprocals   – **Up to 30th**

Tables 2nd to 30th

S.No	Table			
1	2	3	4	5
2	4	6	8	10
3	6	9	12	15
4	8	12	16	20
5	10	15	20	25
6	12	18	24	30
7	14	21	28	35

8	16	24	32	40
9	18	27	36	45
10	20	30	40	50

6	7	8	9	10
12	14	16	18	20
18	21	24	27	30
24	28	32	36	40
30	35	40	45	50
36	42	48	54	60
42	49	56	63	70
48	56	64	72	80
54	63	72	81	90
60	70	80	90	100

11	12	13	14	15
22	24	26	28	30
33	36	39	42	45
44	48	52	56	60
55	60	65	70	75
66	72	78	84	90
77	84	91	98	105
88	96	104	112	120
99	108	117	126	135
110	120	130	140	150

16	17	18	19	20
32	34	36	38	40
48	51	54	57	60
64	68	72	76	80
80	85	90	95	100
96	102	108	114	120
112	119	126	133	140
128	136	144	152	160
144	153	162	171	180
160	170	180	190	200

21	22	23	24	25
42	44	46	48	50
63	66	69	72	75
84	88	92	96	100
105	110	115	120	125
126	132	138	144	150
147	154	161	168	175
168	176	184	192	200
189	198	207	216	225
210	220	230	240	250

26	27	28	29	30
52	54	56	58	60
78	81	84	87	90

104	108	112	116	120
130	135	140	145	150
156	162	168	174	180
182	189	196	203	210
208	216	224	232	240
234	243	252	261	270
260	270	280	290	300

Squares 1 to 40

No.		Square
1	→	1
2	→	4
3	→	9
4	→	16
5	→	25
6	→	36
7	→	49
8	→	64
9	→	81
10	→	100

No.		Square
11	→	121
12	→	144

13	→	169
14	→	196
15	→	225
16	→	256
17	→	289
18	→	324
19	→	361
20	→	400

No.		Square
21	→	441
22	→	484
23	→	529
24	→	576
25	→	625
26	→	676
27	→	729
28	→	784
29	→	841
30	→	900

No.		Square
31	→	961
32	→	1024
33	→	1089

34	→	1156
35	→	1225
36	→	1296
37	→	1369
38	→	1444
39	→	1521
40	→	1600

Cube 1 to 25

No.		Cube
1	→	1
2	→	8
3	→	27
4	→	64
5	→	125
6	→	216
7	→	343
8	→	512
9	→	729
10	→	1000

No.		Cube
11	→	1331
12	→	1728

No.		Cube
13	→	2197
14	→	2744
15	→	3375
16	→	4096
17	→	4913
18	→	5832
19	→	6859
20	→	8000

No.		Cube
21	→	9261
22	→	10648
23	→	12167
24	→	13824
25	→	15625

Square Roots 1 to 20

No.		Square Root
1	→	1
2	→	1.414

No.		Square Root
11	→	3.3166
12	→	3.464

No.		Square Root		No.		Square Root
3	→	1.732		13	→	3.6056
4	→	2		14	→	3.742
5	→	2.236		15	→	3.873
6	→	2.449		16	→	4
7	→	2.6458		17	→	4.123
8	→	2.828		18	→	4.243
9	→	3		19	→	4.359
10	→	3.162		20	→	4.472

Cube Roots 1 to 20

No.		Cube Root		No.		Cube Root
1	→	1		11	→	2.224
2	→	1.26		12	→	2.29
3	→	1.44		13	→	2.35
4	→	1.5874		14	→	2.41
5	→	1.709		15	→	2.47
6	→	1.82		16	→	2.52
7	→	1.913		17	→	2.57
8	→	2		18	→	2.62
9	→	2.08		19	→	2.621
10	→	2.1544		20	→	2.714

Reciprocals 1 to 30 (Reciprocal % Equivalent)

No.		Reciprocal
1/1	→	100%
1/2	→	50%
1/3	→	33.33%
1/4	→	25%
1/5	→	20%
1/6	→	16.67%
1/7	→	14.29%
1/8	→	12.50%
1/9	→	11.11%
1/10	→	10.00%

No.		Reciprocal
1/11	→	9.0909%
1/12	→	8.3333%
1/13	→	7.6920%
1/14	→	7.1430%
1/15	→	6.6700%
1/16	→	6.25%
1/17	→	5.88%
1/18	→	5.56%
1/19	→	5.26%
1/20	→	5%

No.		Reciprocal
1/21	→	4.762%
1/22	→	4.5455%
1/23	→	4.3478%
1/24	→	4.1670%
1/25	→	4%
1/26	→	3.85%
1/27	→	3.70%
1/28	→	3.57143%
1/29	→	3.4483%
1/30	→	3.33%

Exercise – 1

1. Fill up following blanks on the basis of tables

   a) 9 X ☐ = 81
   b) 12 X 6 = ☐
   c) ☐ ☐ 9 = 63
   d) ☐ X 5 = 65
   e) 19 X 9 = ☐
   f) 21 X 9 = ☐
   g) 18 X ☐ = 90
   h) 23 X 3 = ☐
   i) 5 X ☐ = 35
   j) 14 X ☐ = 84

2. Fill what is next in blank space on the basis of square

   a) 2^2 = _____
   b) 4^2 = _____
   c) _____ = 81
   d) $5^☐$ = 25
   e) $☐^2$ = 36
   f) $☐^2$ = 64
   g) $☐^2$ = 81
   h) 13^2 = _____
   i) $1☐^2$ = 256
   j) 17^2 = _____
   k) 18^2 = _____
   l) $☐^2$ = 625
   m) $1☐^2$ = 121

3. Fill appropriate natural number on the basis of cube rule or vice versa.

a) $1000 = \square^3$
b) $27 = \square^3$
c) $125 = \square^3$
d) $\square = 3^3$
e) $3375 = (\_\_)^3$
f) $5832 = (\_\_)^3$
g) $9261 = (2\_)^3$
h) $6^3 = \_\_\_\_$
i) $9^3 = \_\_\_\_$
j) $6^3 - 1 = \square$
k) $6^3 - 1 = \square$
l) $64 - 4^3 = \square$
m) $4913 - 16^3 = \square$
n) $24^3 - 21^3 = \square$
o) $512 - 13^3 = \square$

4. Fill the answers on the basis of square roots or vice versa.

a) $6 = \square$
b) $19 = \square$
c) $15 = \square$
d) $16 = \square$
e) $3.873 = \square$
f) $3.742 = \square$
g) $12 = \square$
h) $16 = \square$
i) $\sqrt{3} = \_\_\_\_$
j) $\sqrt{2} = \_\_\_\_$

5. Fill the answers on the basis of cube roots or vice versa
   a) $\sqrt[3]{3}$ = ☐
   b) $\sqrt[3]{2}$ = ☐
   c) $2^{1/2}$ = ☐
   d) $9^{1/3}$ = ☐
   e) 2.1544 = ☐
   f) $\sqrt[3]{13}$ = ☐
   g) 2.621 = ☐
   h) $\sqrt[3]{18}$ = ☐
   i) $\sqrt[3]{11}$ = ☐
   j) $11^{1/3}$ = ☐
   k) $18^{1/3}$ = ☐
   l) $6^{1/3}$ = ☐

6. Fill the appropriate answers on the basis of reciprocals
   a) 2 = ☐
   b) 6 = ☐
   c) 5 = ☐
   d) 10 = ☐
   e) 12 = ☐
   f) 11 = ☐
   g) 12 = ☐
   h) 15 = ☐
   i) 17 = ☐
   j) 18 = ☐

Test Yourself (9.1) Time 10Min.

1. $3^2 \times 4^2 =$ ☐
2. $4^2 \times 2^3 =$ ☐
3. $6^3 + 7^2 =$ ☐
4. $1/10 =$ ☐ - in %
5. $26^2 =$ ☐
6. $19^3 =$ ☐
7. $3^3 - 3^3 =$ ☐
8. $1/19 =$ ☐
9. $\sqrt{15} =$ ☐
10. $15^{1/2} =$ ☐
11. $13^{1/2} =$ ☐
12. $13^{1/3} =$ ☐
13. $1/9 =$ ☐ - in %
14. $20^3 =$ ☐
15. $27^2 + 3^3 + 0.01 =$ ☐

Note : Each question having one number each.

If you score
15/15= If this is your score than you can move on.

(10-14)/15= If your score lie between them, than revise basic tools and attempt test 9.2.
(1-10)/15= If your score lie between them, than learnt by heart every day until you feel confident. Once you confident then attempt test 9.1 once again & follow instruction according.

Test Yourself (9.2)   Time 10Min.

1. What will be the answer if we subtract 1 from square root of **25**
2. 23% of 425 = ☐
3. 5% of 1200 = ☐
4. 12.5% of 100 = ☐
5. $4^3 - 1$ = ☐
6. $6^3 + 1 - 0.02$ = ☐
7. $\sqrt{2}$ = ☐
8. $\sqrt{4^2}$ = ☐
9. $\sqrt{25} - 1$ = ☐
10. $12 \times 12 + 1 - 1$ = ☐
11. $325.04 + 0.01$ = ☐
12. $\sqrt[3]{10}$ = ☐
13. 28×7 = ☐
14. $24 \times$ ☐ = 240
15. $13 \times$ ☐ $- 1 = 116$

Note :- Each Question having one number each If you Score

15/15 = If this is your score than you can move on
10-14/15 = If your score lies between them, then revise basics tools and do more practice.
1-10/15 = If your score lies between them, then learnt by heart every day and memorize before sleep. Once you confident then rest for two day and attempt once again test 9.1 and follow instruction.

Learning result is really useful?
- If you crossed test yourself 9.2 with honesty. I can assure you; you are 10 times more confident among your age group guys.

- Avoid using the pencils to do a calculation, make a natural habit to calculate orally in day-to-day life.

- This will give you best possible answer without delay which will be boon for competition whether they are student or job preparation guys.
-

Assignment
Start with simple day to day calculations.
1. When you and your close one buy something from the market try to calculate before their outcome.
2. Review financial data in newspaper & discuss with your friends.
3. Start from small calculations and slowly start big.

> Om chant is the only source of natural food for mind/brain.
> ➢ Do practice before sleep

Answer Key

Test Yourself 1.1

a) 10   b) 0.1   c) 3   d) 7
e) 28   f) 140   g) 98   h) 87
i) 65.16   j) 300.96

Test Yourself 2.1

1. 7290   2. 9702   3. 17000   4. 33276
5. 12432   6. 7832   7. 5700   8. 7308
9. 12992   10. 11000

Test Yourself 3.1

1. Base – 100, 9118
2. Base – 100, 8178
3. Base – 100, 5922
4. Base – 100, 11130
5. Base – 100, 10656
6. Base – 200, 38190
7. Base – 200, 39457
8. Base – 300, 71337
9. Base – 300, 73416
10. Base – 300, 112200
11. Base – 400, 193920
12. Base – 500, 234901
13. Base – 600, 409725
14. Base – 700, 543976
15. Base – 800, 632748
16. Base – 600, 416070
17. Base – 800, 649611
18. Base – 900, 816306
19. Base – 300, 98568
20. Base – 800, 622377

Test Yourself 4.1

1. 1216
2. 11009
3. 497009
4. 616
5. 366021
6. 366628
7. 93021
8. 9212
9. 164009
10. 366024

Test Yourself 5.1

1. 2915
2. 2475
3. 102375
4. 384375
5. 44075
6. 372075
7. 9975
8. 10605
9. 12075
10. 3575

Test Yourself 7.1

1. 2601
2. 3969
3. 1156
4. 6241
5. 13456
6. 96100
7. 190096
8. 561001
9. 404496
10. 58081
11. 14161
12. 121801
13. 541696
14. 678976
15. 904401
16. 110889
17. 1764
18. 11881
19. 40804
20. 7744

Test Yourself 8.1

1. b
2. c
3. c
4. a
5. a

Test Yourself 9.1

1. 25
2. 24
3. 265
4. 10%
5. 676
6. 6859
7. 0
8. 0.053
9. 3.873
10. 3.873
11. 3.6056
12. 2.35
13. 11.111%
14. 8000
15. 756.01

Test Yourself 9.2

1. 4
2. 97.75
3. 60
4. 12.5
5. 63
6. 216.98
7. 1.414
8. 4
9. 4
10. 144
11. 325.05
12. 2.154
13. 196
14. 10
15. 9

SOME COMPETITIVE EXAMS' EXAMPLE FOR GENERAL PRACTICE

A) A student who scored 30% marks in the first paper of physics out of 180 marks, has to get an overall score of at least 50% in the paper. The second paper is of 150 marks. The percentage of marks he should score in the second paper to get the overall average score is

(1) 80%   (2) 76%   (3) 74%   (4) 70%

B) Simplify

$$\sqrt{-\sqrt{3}+\sqrt{\sqrt{(3+8\sqrt{7}+4\sqrt{3})}}}$$

(1) $\sqrt{3}$ (2) 5   (3) $\sqrt{2}$ (4) 2

C) Simplify $\dfrac{0.035 \times 0.035 \times 0.035 + (0.965)^3}{(0.035)^2 - 0.035 \times 0.965 + (0.965)^2}$

(1) 5   (2) 0.5   (3) 1   (4) 0.4

D) A student has to get 40% marks to pass in an examination suppose he gets 30 marks and fails by 30 marks then what one the maximum marks in the examination?

(C-SAT 2018, UPSC)

(1) 100   (2) 120   (3) 150   (4) 300

E) A train 200 meters long is moving at the rate of 40 KMPH. In how many seconds will it cross as man standing near the railway line?

(C-SAT 2018, UPSC)

(1) 12 (2) 15 (3) 16 (4) 18

F) Graph based questions
Directions

The following three items one based on the graph given below which shows imports of three different types of steel over a period of six months of a year. Study the graph and answer the three items that follows

(C-SAT 2018, UPSC)

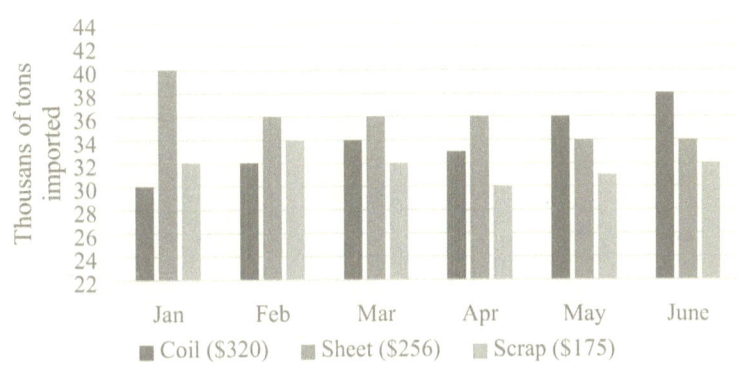

The figure in the brackets indicate the average cost per ton over six months period.

(F1) By how much (measured in thousands of tons) did the import of sheet steel exceed the import of coil steel in the first 3 months of the year?
(1) 11 (2) 15 (3) 19 (4) 23

(F2) What was the approximate total value (in $) of sheet steel imported over the six months period?
(1) 45555 (2) 50,555 (3) 55,550 (4) 65750

(F3) What was the approximate ratio of sheet steel and scrap steel imports in the first 3 months of the year?

(1) 1:1 (2) 1.2:1 (3) 1.4:1   (4) 1.6:1

(G) A shopkeeper sells an article at ₹ 40 and gets X% profit. However, when he sells it at ₹ 20, he faces same % of loss. What is the original cost of the article?

(C-SAT 2018, UPSC)

(1) ₹ 10   (2) ₹ 20   (3) ₹ 30   (4)   ₹ 40